KU-570-965

FOR AGES 7-9
BRILLIANT
IDEAS FOR
TIMES
TABLES
Practice

by Molly Pot

ANDREW BRODIE

7000000260246

First published 2012 by Andrew Brodie
an imprint of Bloomsbury Publishing Plc
50 Bedford Square
London WC1B 3DP
ISBN 9781408181317

Text © Molly Potter 2012
Illustrations © Mike Phillips/Beehive Illustration

A CIP record for this publication is available from the British Library.
All rights reserved. This book may be photocopied, for use in the educational establishment for which
it was purchased, but may not reproduced in any other form or by any means - graphic,
electronic, or mechanical, including photocopying, recording, taping or information storage or
retrieval systems - without the prior permission in writing of the publishers.

Printed and bound by CPI Group (UK) Ltd, Croydon CR0 4YY

1 3 5 7 9 10 8 6 4 2

This book is produced using paper that is made from wood grown in
managed, sustainable forests. It is natural, renewable and recyclable.
The logging and manufacturing processes conform to the environmental
regulations of the country of origin.

To see our full range of titles visit www.bloomsbury.com

Contents

Introduction

THIS BOOK

This book provides a selection of activities to help pupils not only to learn their tables, but also to feel more positive about them. It includes a mix of straight forward and more involved and creative activities to encourage pupils to associate tables with fun while they learn them for life.

HOW TO USE THIS SERIES

This book and the others in the series support the learning of each multiplication table starting with a poster that can be used to introduce each specific table. This can be coloured in and used by the pupils at home to aid their learning. The easier activities start each section dedicated to a particular table. Many of these can be used more than once to teach and revisit each table – especially the revision activities. Some more difficult sheets could be enlarged to A3 and pupils can work together to solve the challenges.

NEW APPROACHES TO LEARNING

Why not decide to make this the year that all of your class learn their tables once and for all? What a gift to give them! With a little extra effort this is quite possible. Here are some suggestions to help you:

• *Tackle the motivation to learn multiplication tables and find the best ways to learn them*
Discuss why times tables are in the curriculum and how knowing them not only makes their school maths career much easier but stands them in good stead for the future. For example, as an adult they might need to find out the cost of several items of the same price, the amount needed for a meal, how to calculate how many days there are in a number of weeks and so on.

Have a discussion about learning tables – what makes it hard, what stops some people from learning them? Ask pupils to discuss in pairs what they believe would help them to learn their multiplication tables or if they have already mastered them – what did they do? Ask pupils to write their ideas down on sticky notes and stick them onto a large sheet of paper (for younger children, just pool your ideas). Discuss the ideas they create and make a plan to help your class learn their tables.

• *Create a 'Table of the week' each week*
Choose one of the more difficult tables to focus on each week. Display that table with its answer and its inverse sum boldly in the classroom. For example $8 \times 7 = 56$ and $7 \times 8 = 56$. If possible, include an interesting fact about the answer or a picture that relates to the sum or answer.

• *List difficult tables*
Some tables are definitely more difficult to remember than others. Create a 'tricky tables' poster. Pupils could agree on a mark out of ten for the difficulty of each sum.

• *Have individual table cards for pupils*
Create differentiated lists of multiplication tables with three columns beside each sum. Pupils can colour inside the three columns next to each table red (do not know it), then add amber (nearly know it) and finally green (always get it right). Allow pupils time to look at this card regularly to review their own learning and try to learn those they have not yet mastered.

• *Play tables games*
Use games such as those included in these books (especially 9–11) regularly to revise the tables in an entertaining way.

A whole-class game that can be used to revise tables is 'winner stays on'. First choose two pupils to stand up. Next ask a times table question. The first pupil to shout out the answer stays standing, the other pupil sits down. Another pupil is asked to stand, another sum issued and again the first person to shout the answer is the winner who then stays on and so on.

• *Create a times tables test and award*
To achieve this award any pupil has to correctly answer 20 random multiplication table questions of those that have been covered in your class. If they succeed they can be issued the award (a badge, a certificate or a significant number of 'merits' from the existing classroom reward system). Explain to the class that they can request to take the tables test at any point in the year to try and achieve the award to aid continuous motivation to learn.

Brilliant Ideas for Times Tables 7–9 © Molly Potter 2012

Which is the correct answer?

x2
x5
x10

Name _____ **Date** _____

Circle the correct answer for each of these sums.

①	4 x 2 =	2	5	8	10
②	6 x 5 =	20	25	30	40
③	7 x 10 =	40	50	60	70
④	2 x 8 =	12	14	16	18
⑤	5 x 5 =	15	20	25	50
⑥	9 x 10 =	90	70	80	100
⑦	6 x 2 =	10	12	14	16
⑧	7 x 5 =	25	35	30	40
⑨	10 x 10 =	20	50	100	120
⑩	9 x 5 =	30	35	40	45
⑪	9 x 2 =	12	14	16	18
⑫	3 x 5 =	5	10	15	25
⑬	2 x 10 =	30	20	10	5
⑭	4 x 5 =	10	20	30	40
⑮	2 x 3 =	2	4	6	8
⑯	10 x 3 =	10	30	50	15
⑰	2 x 5 =	5	15	10	20
⑱	1 x 10 =	5	12	15	10
⑲	2 x 2 =	1	2	4	6
⑳	5 x 8 =	30	20	16	40

Colour by numbers

Name _____ **Date** _____

Work out these sums and colour in the shapes that contain the answers. What picture appears?

3 x 5 =	2 x 5 =	7 x 10 =	2 x 2 =	6 x 5 =
8 x 10 =	5 x 10 =	8 x 5 =	6 x 10 =	2 x 3 =
2 x 6 =	2 x 7 =	2 x 9 =	2 x 10 =	10 x 6 =
9 x 10 =	1 x 5 =	5 x 7 =	2 x 4=	2 x 2 =
8 x 2 =	5 x 5 =	10 x 10 =	9 x 5 =	

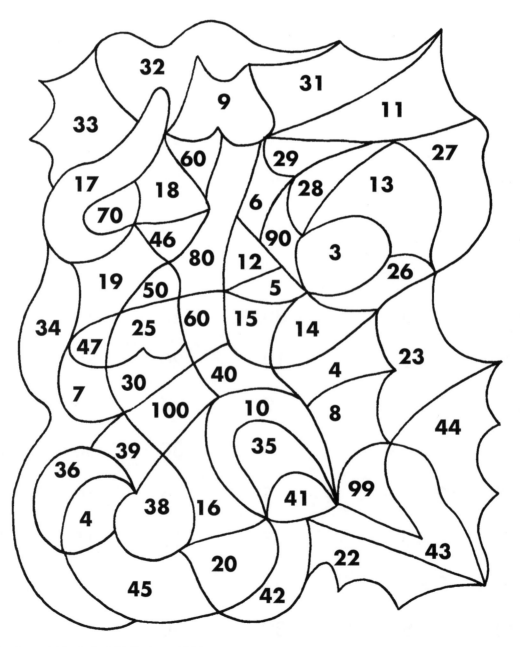

Which is closer?

Name _____ **Date** _____

For each pair of sums, circle the one that has its answer closer to it.

For example – with the first pair of sums (2 x 8 or 4 x 10) the answer 16 is closer than the answer 40, so you need to circle 2 x 8.

(2 x 8) or 4 x 10

9 x 5 or 2 x 7

8 x 10 or 7 x 5

5 x 5 or 9 x 2

10 x 10 or 5 x 5

8 x 2 or 8 x 5

2 x 2 or 5 x 10

2 x 7 or 6 x 5

10 x 7 or 2 x 4

9 x 5 or 2 x 3

3 x 5 or 2 x 6

9 x 10 or 5 x 5

45 16 30 8 35 100 40 50 18 12 14 80 6 70 15 4 90 25

Which sum has the most answers?

x2 x5 x10

Name _____ **Date** _____

Tally the answers to the sums to find out which sum has its answer written the most times. Cross out the answers after you have counted them. One has been done for you.

Sum (2 x)	Number of answers	Sum (5 x)	Number of answers	Sum (10 x)	Number of answers
2 x 3	1	5 x 3		10 x 4	
2 x 4		5 x 4		10 x 7	
2 x 5		5 x 5		10 x 9	
2 x 6		5 x 6		10 x 10	
2 x 7		5 x 7			
2 x 8		5 x 9			
2 x 9					

70 100 40 90
14 15 14 16 15
35 18
45 10 18
18 12 30 45 14
20 70 12
15 35 6 35
45
25 10 18 16 18
14 90 25
8 30 45 15 20 40
18 8
16 12 14 30
35 100

Find out about Snook

x2
x5
x10

Name _____ **Date** _____

Use this de-coder to find out about Snook!

2	4	5	6	8	10	12	14	15	16	18	20	25	30	35	40	45	50	60	70	80	90	100
a	b	c	d	e	f	g	h	i	j	k	l	m	n	o	p	r	s	t	u	v	w	y

What does Snook like to eat?

___ ___ ___ ___ ___   ___ ___ ___ ___
(1 x 4) (9 x 5) (3 x 5) (5 x 1) (9 x 2)   (5 x 10) (10 x 6) (4 x 2) (9 x 10)

What does Snook love to drink?

___ ___ ___ ___ ___ ___ ___
(2 x 3) (10 x 7) (5 x 10) (10 x 6) (2 x 2) (3 x 5) (6 x 5)

___ ___ ___ ___ ___
(2 x 8) (7 x 10) (5 x 3) (1 x 5) (2 x 4)

What is Snook's favourite toy?

___ ___ ___   ___ ___ ___ ___
(6 x 10) (7 x 2) (2 x 4)   (5 x 5) (7 x 5) (5 x 7) (6 x 5)

Where does Snook live?

___ ___   ___   ___ ___ ___ ___
(3 x 5) (5 x 6)   (1 x 2)   (10 x 5) (7 x 5) (5 x 1) (2 x 9)

What does Snook always attack?

___ ___ ___ ___ ___ ___ ___
(6 x 10) (9 x 5) (2 x 1) (2 x 5) (1 x 10) (3 x 5) (5 x 1)

___ ___ ___ ___ ___ ___
(4 x 5) (5 x 3) (2 x 6) (7 x 2) (10 x 6) (10 x 5)

Now use the de-coder to make up the answers to these two questions about Snook. You will need more paper to do this. See if your friend can work out the questions!

1) What is Snook's last name?   2) What scares Snook?

Play bingo with a friend

x2
x5
x10

Name _____ **Date** _____

- Find someone to play with and each choose a different bingo card from this page.

- Cut out the sum cards from page 11 and shuffle them. Place them in a pile face down.

- Take the cards one at a time and read the question. If you have the answer to the sum on your bingo card, cross it out.

- The winner is the first person to cross out four in a row – up, across or diagonally.

- Make sure you both get your sums right! Play again with different bingo cards.

25	35	12	70
30	14	90	2
8	10	6	18
20	16	100	45

4	70	12	100
60	15	80	8
14	18	6	35
20	16	40	25

35	70	14	5
10	25	18	16
15	8	60	6
40	20	2	50

90	14	2	30
70	5	10	60
20	8	35	6
18	45	80	12

12	80	45	50
2	5	10	14
20	30	16	90
6	35	100	4

16	15	50	4
45	80	12	20
35	40	70	8
6	60	25	90

 Brilliant Ideas for Times Tables 7–9 © Molly Potter 2012

1 x 2 =	1 x 5 =	1 x 10 =
2 x 2 =	2 x 5 =	2 x 10 =
3 x 2 =	3 x 5 =	3 x 10 =
4 x 2 =	4 x 5 =	4 x 10 =
5 x 2 =	5 x 5 =	5 x 10 =
6 x 2 =	6 x 5 =	6 x 10 =
7 x 2 =	7 x 5 =	7 x 10 =
8 x 2 =	8 x 5 =	8 x 10 =
9 x 2 =	9 x 5 =	9 x 10 =
10 x 2 =	5 x 10 =	10 x 10 =

Dot to dot

Name _____   **Date** _____

You need to join all the sums to their answers with straight lines.
All the sums are in rectangles and all the answers are in circles.
One sum (3 x 5 = 15) has been done for you. Cross out each sum
and its answer after you have joined them. When you have joined
all the sums to their answers, you will have written two words.
What does it say?

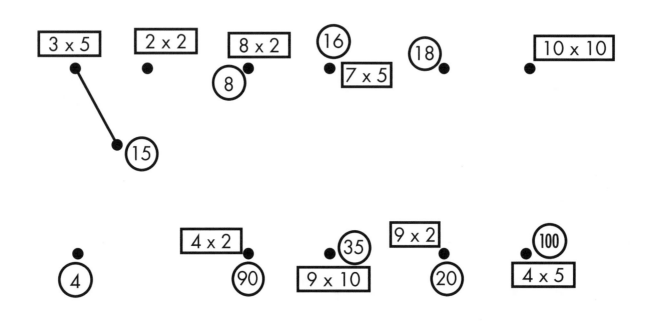

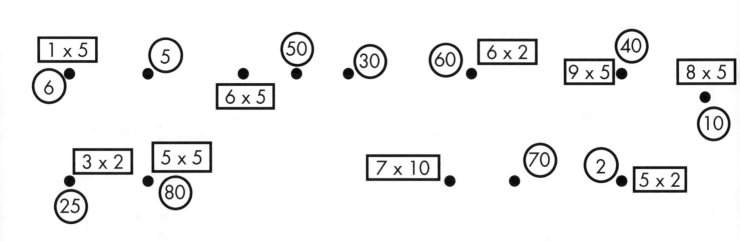

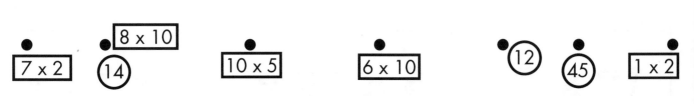

Multiply by ?

Name _____ **Date** _____

Multiply the numbers in each table by the number above it and put the answer in the space below. One has been done for you.

x 2

5	2	8	4	1	10	7	3	6	9
10									

x 5

10	3	2	8	5	1	7	6	4	9

x 10

5	1	4	7	10	9	3	2	8	6

x 2

6	5	10	2	1	4	3	7	8	9

x 5

3	8	5	1	7	2	8	4	6	10

How many?

Name _____   **Date** _____

Work out the following sums and then write the answers. The first one
has been done for you.

Fooz

Groot

Harf

How many ...

(1) Legs would two Fooz have?

    2 x 8 = 16

(2) Feet would two Groot have?

_____

(3) Legs would ten Harf have?

_____

(4) Eyes would five Fooz have?

_____

(5) Antennae would two Fooz have?

_____

(6) Teeth would five Groot have?

_____

(7) Eyes would five Groot have?

_____

(8) Ears would five Groot have?

_____

(9) Legs would ten Fooz have?

_____

(10) Eyes would ten Groot have?

_____

(11) Spikes on the back would six
Harf have?

_____

(12) Legs would five Fooz have?

_____

(13) Antennae would five Fooz have?

_____

(14) Legs would five Groot have?

_____

The 3 x Table

Match it

Name _____ **Date** _____

Match each multiplication sum to its answer. Use colours and patterns to make each sum look the same as its answer. One has been done for you.

24

 x 3

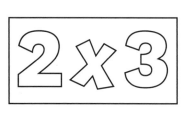 2 x 3   9

15

21

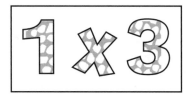 3 x 3

21

 3

 10 x 3

27

 4 x 3

6   12

18   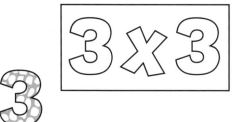 5 x 3

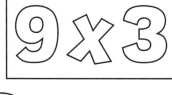 9 x 3   12   27

18   24    6 x 3

30   15   9

6   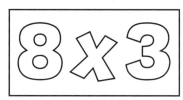 8 x 3    7 x 3   3

Get rid of the wrong answers

Name _____ **Date** _____

Answer the sum in each of these shapes and then colour the wrong numbers black. One has been done for you. When you have finished, a picture will appear.

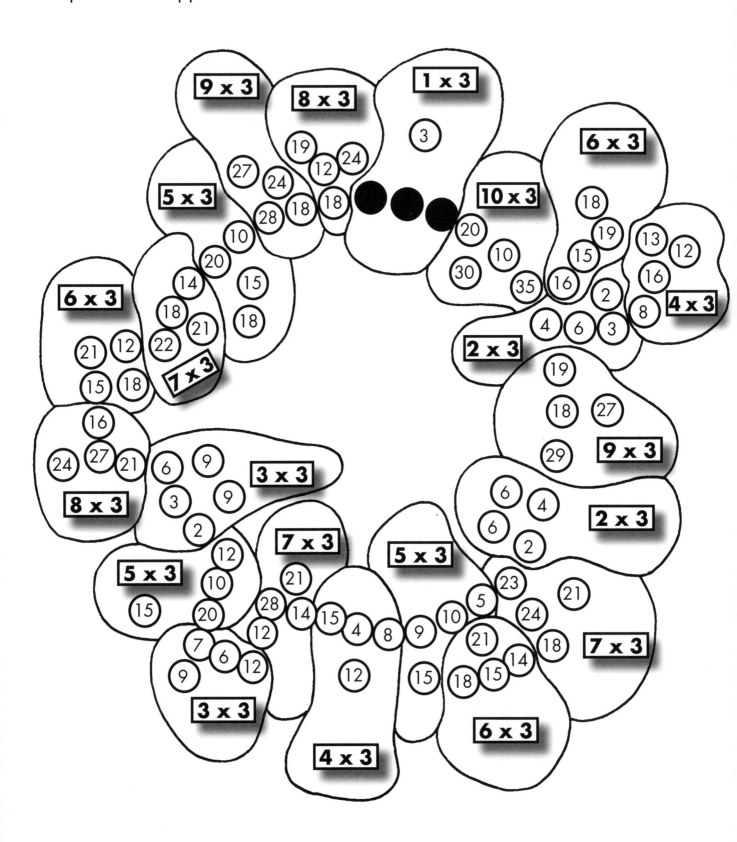

Alien grub

Name _____ **Date** _____

This is a bowl of alien food. If you were an alien you would love it!
However the picture is not complete – it needs some colours and patterns.
Can you use the tables key to complete the picture and then label the different
foods? One has been done for you.

Answer	Name of food	Colour/pattern
3	spotted gudge	white with black spots
6	grox	blue with green triangles
9	spok stick	brown and white stripes
12	triggle	yellow
15	fripple	black and white stripes
18	tuff	blue
21	yock	big red and yellow blobs
24	kitter	green with little purple bits
27	pilp	red and purple scribble
30	edible dish!	orange

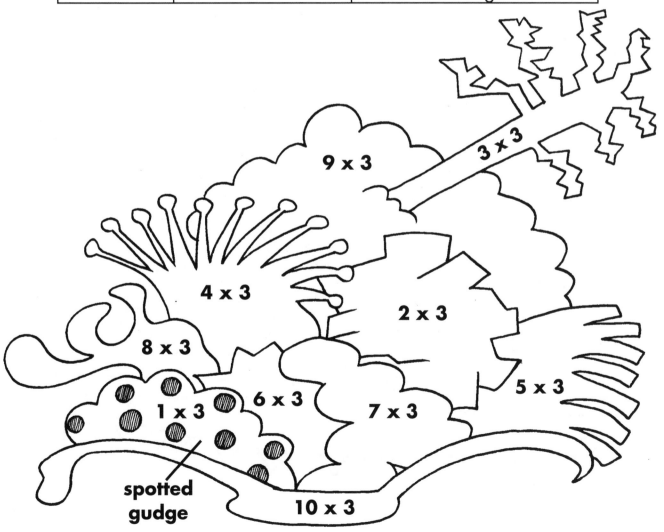

Label the picture

Name _____ **Date** _____

Can you use sums from the three times table to label the objects in this picture? One has been done for you. See how many items you can label.

1 x 3 drips

HAPPY BIRTHDAY

Brilliant Ideas for Times Tables 7–9 © Molly Potter 2012

Complete the picture

Name _____ **Date** _____

Complete this picture by first doing the multiplication sums and then by drawing in that number of the thing described to the picture of the alien.

8 x 3 hairs on both hands in total

10 x 3 hairs on head

4 x 3 teeth sticking out

2 x 3 eyes

9 x 2 spots on his face

1 x 3 fingers on each hand

5 x 3 stars like this ✳ in the pattern on his top

7 x 3 buttons on its top

6 x 3 is the number on his pendant

3 x 3 very thin legs

Monster sandwich

Name _____ **Date** _____

OK brace yourself! Uggle the monster wants you to make him a sandwich. Work out the sums to find out just how Uggle wants his sandwich to be made and then draw the layers.

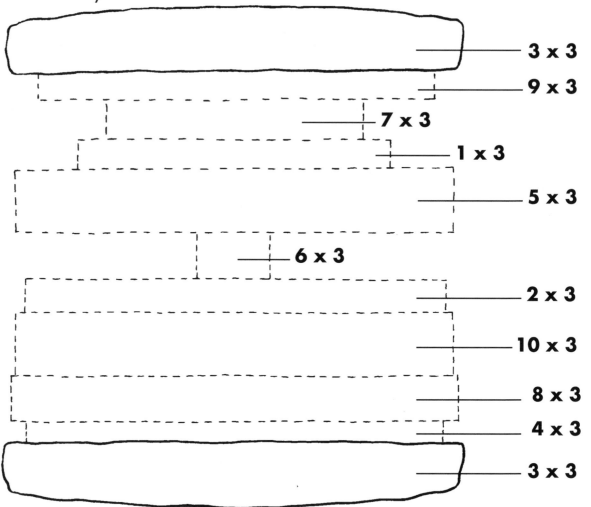

3 x 3
9 x 3
7 x 3
1 x 3
5 x 3
6 x 3
2 x 3
10 x 3
8 x 3
4 x 3
3 x 3

Answer	Layer
3	a big fat caterpillar
6	worms
9	bread with bogey bits in
12	a layer of leaves to stop the bread going soggy
15	some litter
18	an eyeball
21	mayonnaise with ants in
24	different sized spiders
27	a layer of grass to stop the bread going soggy
30	feathers

Which is the right answer?

x3   **Name** _____   **Date** _____

Circle the correct answer for each sum.

1 x 3 = 1 6 3 9 0

3 x 1 = 1 2 4 3

2 x 3 = 0 2 3 9 6

3 x 4 = 4 6 16 12 9

3 x 3 = 2 6 4 3 9 15

3 x 8 = 10 12 18 16 24

4 x 3 = 6 15 12 8 9

3 x 9 = 18 20 27 6

5 x 3 = 5 10 9 30 15

3 x 6 = 27 24 12 16 15 18

6 x 3 = 6 10 12 30 18

3 x 5 = 5 15 20 10

7 x 3 = 14 24 42 21 16 18

3 x 2 = 6 12 9 8 20

8 x 3 = 0 10 14 24 21

3 x 7 = 10 21 16 12 24 15

9 x 3 = 6 14 27 18 21

3 x 10 = 9 14 2 10 20 30

10 x 3 = 22 20 14 10 30

3 x 3 = 6 3 9 12 10 22

11 x 3 = 24 18 10 33 11

3 x 11 = 20 30 33 27

12 x 3 = 42 36 22 12 38

3 x 12 = 14 12 36 24 8

Which is closer?

Name _____   **Date** _____

For each pair of sums, circle the one that has its answer closer to it.
For example – with the first pair of sums (1 x 3 or 7 x 3) the answer 3 is closer than the answer 21, so you need to circle 1 x 3.

(1 x 3)　or　7 x 3

2 x 3　or　4 x 3

8 x 3　or　6 x 3

2 x 3　or　5 x 3

3 x 3　or　7 x 3

9 x 3　or　1 x 3

10 x 3　or　4 x 3

6 x 3　or　1 x 3

9 x 3　or　4 x 3

7 x 3　or　5 x 3

6 x 3　or　8 x 3

3 x 3　or　9 x 3

24　　6

15

27

3　　18

21

9

12

30

 Brilliant Ideas for Times Tables 7–9 © Molly Potter 2012

x3

Find out about Jibby

Name _____ **Date** _____

Use this de-coder to find out about Jibby!

0	3	6	9	12	15	18	21	24	27	30
t	r	a	b	o	y	p	s	h	e	c

What is Jibby's favourite thing to eat?

‾‾‾‾ ‾‾‾‾ ‾‾‾‾ ‾‾‾‾
(3 x 8) (2 x 3) (0 x 3) (3 x 7)

What does Jibby want to be when she grows up?

‾‾‾‾　　　　‾‾‾‾ ‾‾‾‾ ‾‾‾‾ ‾‾‾‾ ‾‾‾‾
(2 x 3)　　　　(7 x 3) (8 x 3) (9 x 3) (3 x 9) (6 x 3)

What is Jibby scared of?

‾‾‾‾ ‾‾‾‾ ‾‾‾‾ ‾‾‾‾
(3 x 3) (3 x 4) (3 x 5) (3 x 7)

How does Jibby travel around?

‾‾‾‾ ‾‾‾‾　　　　‾‾‾‾ ‾‾‾‾ ‾‾‾‾ ‾‾‾‾ ‾‾‾‾
(3 x 3) (3 x 5)　　　　(3 x 7) (6 x 3) (2 x 3) (10 x 3) (9 x 3)

‾‾‾‾ ‾‾‾‾ ‾‾‾‾ ‾‾‾‾ ‾‾‾‾ ‾‾‾‾
(3 x 8) (3 x 4) (6 x 3) (3 x 6) (9 x 3) (1 x 3)

What is Jibby's last name?

‾‾‾‾ ‾‾‾‾ ‾‾‾‾ ‾‾‾‾ ‾‾‾‾ ‾‾‾‾ ‾‾‾‾ ‾‾‾‾
(3 x 6) (2 x 3) (3 x 0) (0 x 3) (3 x 4) (3 x 3) (4 x 3) (0 x 3)

Closest to the target

Name _____ **Date** _____

Find out which sum gets closest to the target in the middle of this page.
Answer each sum at the end of the chains and each one will tell you how
many small circles you can move along that line. When you get to the final
circle that's the end of the line and you cannot switch lines.
One example has been done for you.

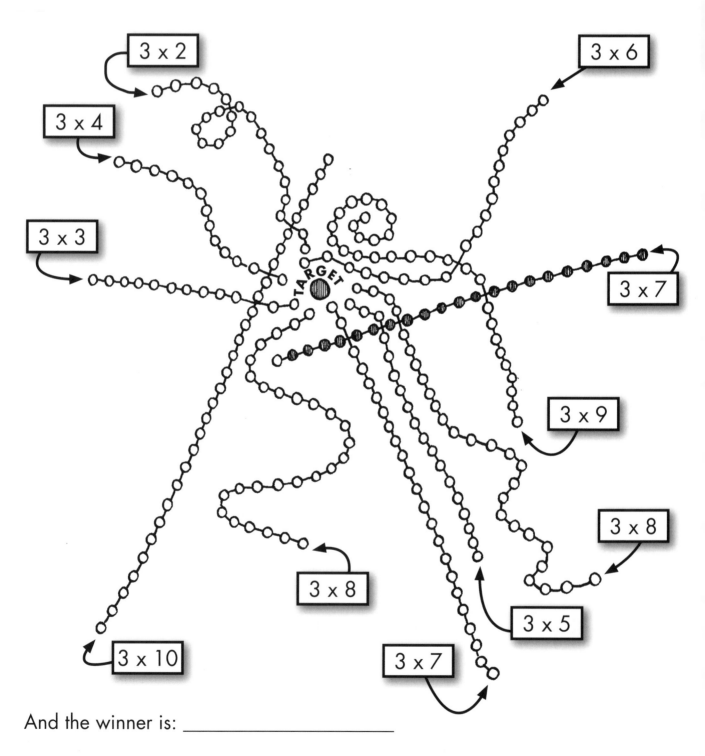

And the winner is: _____

Can you make up a puzzle like this with the two times table?

The 4 x Table

1 x 4 = 4

That's really easy, one four can only ever come to four and four only!

2 x 4 = 8

I can see that!

3 x 4 = 12

This one is really useful. If you have three friends and they all want four cakes you need 12 cakes and a sick bucket!

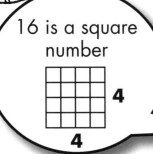

16 is a square number

4 x 4 = 16

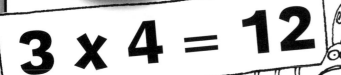

5 x 4 = 20

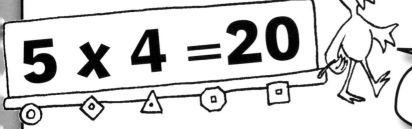

This links to the five times table. I think the five times table is really easy.

6 x 4 = 24

This one rhymes, sort of!

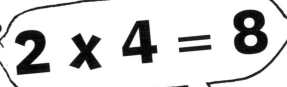

Brilliant Ideas for Times Tables 7–9 © Molly Potter 2012

Join the sum to its answer

Name _____ **Date** _____

Use colouring pencils to join each sum to its answer.
One has been done for you.

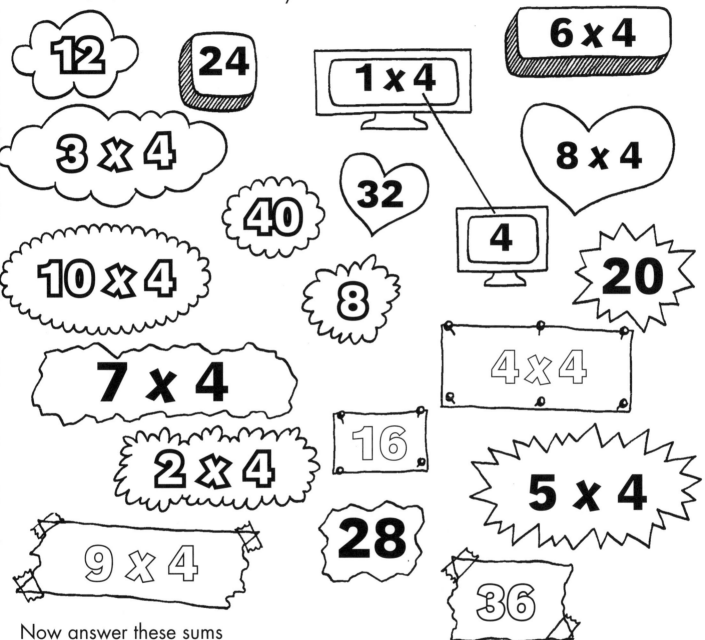

Now answer these sums

5 x 4 =	3 x 4 =	4 x 3 =	4 x 9 =
2 x 4 =	10 x 4 =	1 x 4 =	8 x 4 =
4 x 2 =	4 x 1 =	4 x 5 =	9 x 4 =
4 x 4 =	4 x 2 =	7 x 4 =	4 x 7 =
4 x 6 =	6 x 4 =	4 x 4 =	4 x 10 =

Match it

Name _____ **Date** _____

Match each multiplication sum to its answer. Use colours and patterns to make each sum look the same as its answer. One has been done for you.

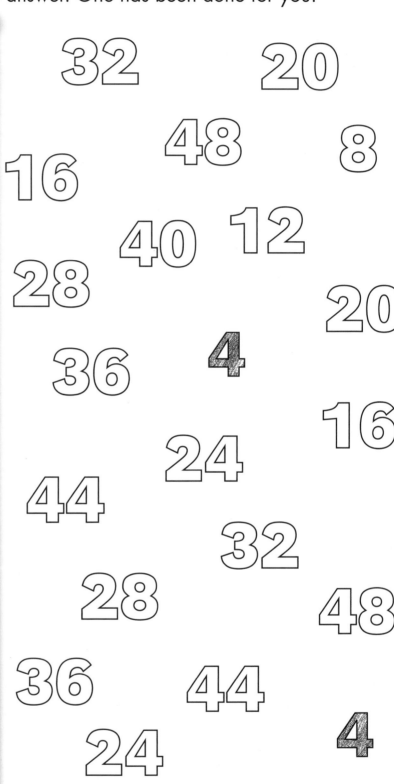

Show the answer

Name _____ **Date** _____

For each of these multiplication sums, colour in the incorrect answers so that all that is left is the one correct answer. One has been done for you.

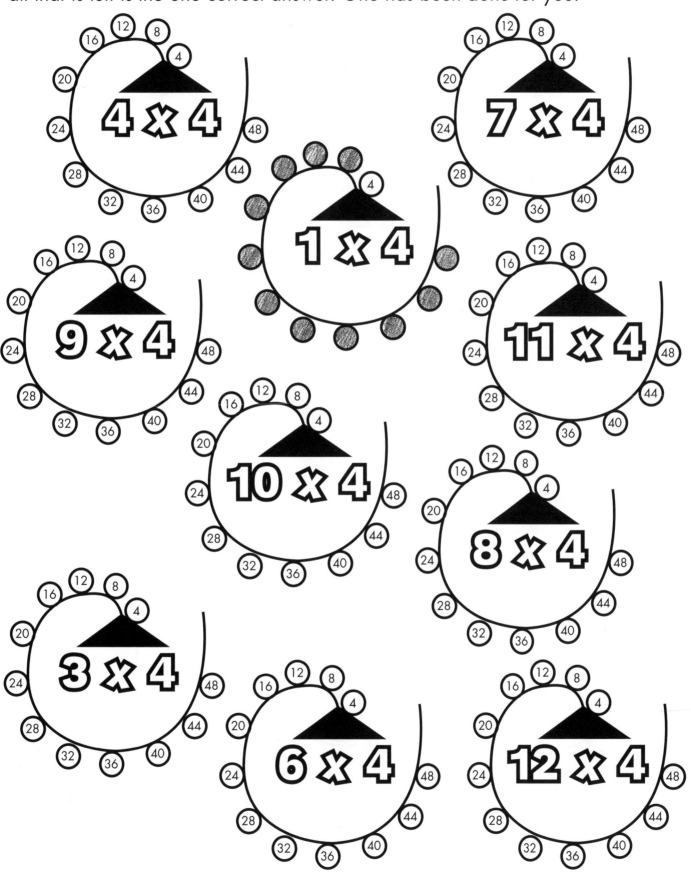

Cast a spell

Name _____ **Date** _____

A naughty witch is about to cast spells on the people and creatures on this page. Use the sums to work out which spell goes where and colour in or draw the person or creature as they will look after the spell.

Sum	Spell
1 x 4	skin turns purple and hair turns green
2 x 4	lots of hair grows on legs
3 x 4	lots of curly hair grows on head
4 x 4	leaves a trail of orange slime behind
5 x 4	flowers grow from ears, clothes, shoes and mouth
6 x 4	becomes short and wide
7 x 4	gets covered in red spots
8 x 4	water spurts from ears
9 x 4	rainbows come from hands
10 x 4	gets tiger stripes

Hidden picture

Name _____ **Date** _____

Colour in all the shapes that contain a number in the four times table and you will see a picture. Can you see what it is?

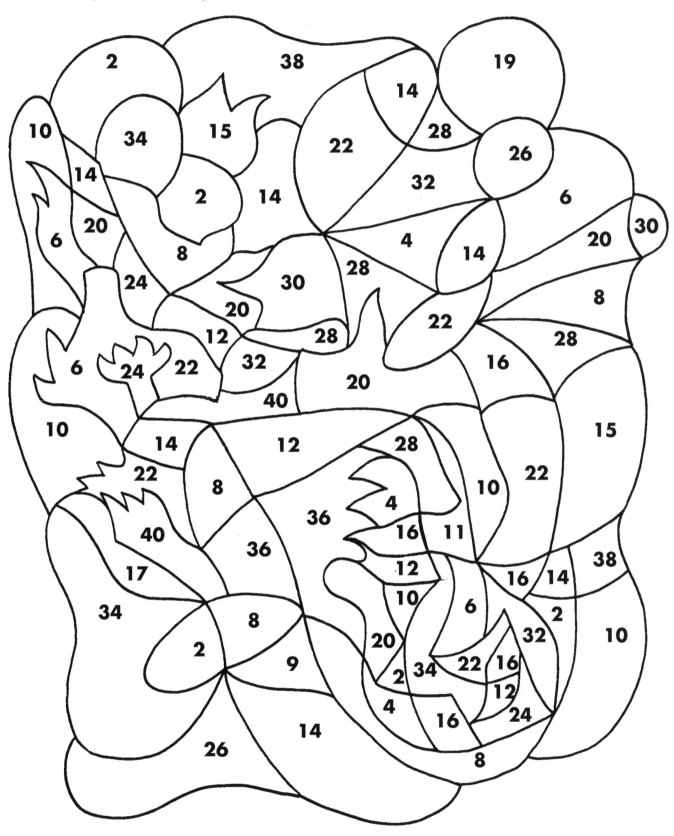

1 x 4   2 x 4   3 x 4   4 x 4   5 x 4   6 x 4   7 x 4   8 x 4   9 x 4   10 x 4

Complete the picture

Name _____ **Date** _____

Complete this picture by adding the things described. One example of each thing is shown but you need to add more of them so you end up with a total amount that is the answer to each sum.

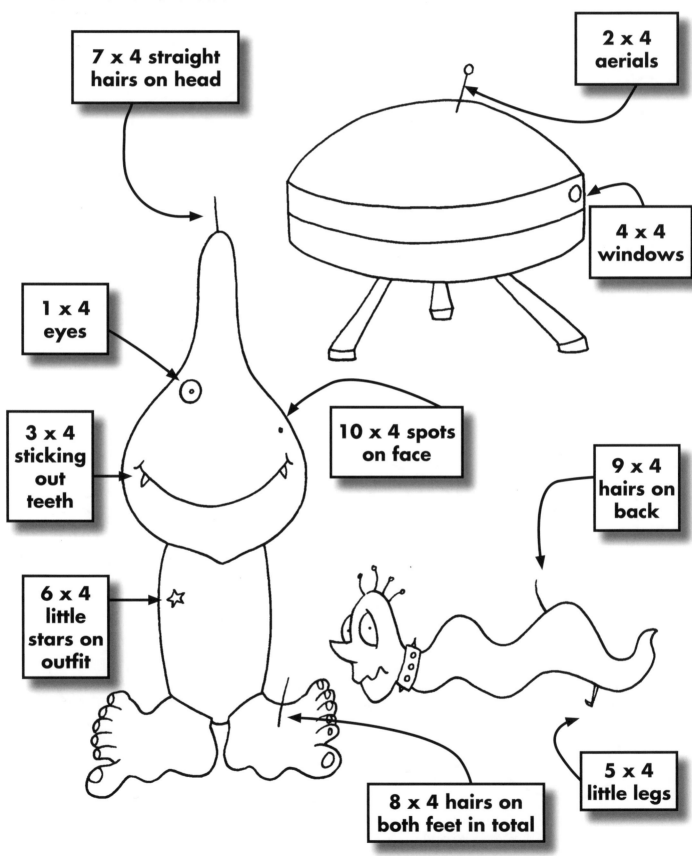

7 x 4 straight hairs on head

2 x 4 aerials

4 x 4 windows

1 x 4 eyes

3 x 4 sticking out teeth

10 x 4 spots on face

9 x 4 hairs on back

6 x 4 little stars on outfit

5 x 4 little legs

8 x 4 hairs on both feet in total

Maze

Name _____ **Date** _____

Use different colours to join each multiplication sum to its answer in the maze.

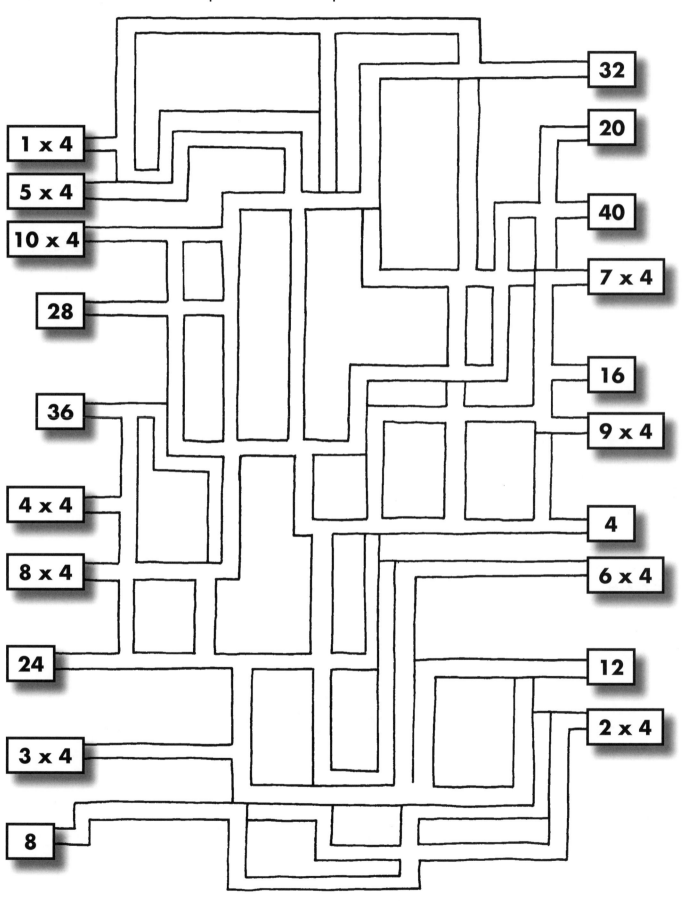

Which is closer?

x4 **Name** _____ **Date** _____

For each pair of sums, circle the one that has its answer closer to it. The first one has been done for you.

6 x 4   or   (5 x 4)

36

7 x 4   or   10 x 4

3 x 4   or   6 x 4

20   **4**   **12**

8 x 4   or   7 x 4

2 x 4   or   5 x 4

28

1 x 4   or   9 x 4

8 x 4   or   4 x 4

9 x 4   or   2 x 4

16

7 x 4   or   3 x 4

40   **24**

8 x 4   or   1 x 4

9 x 4   or   7 x 4

8

32

10 x 4 or   4 x 4

Four times table number search

x4

Name _____ **Date** _____

See how many times you can find the answers to the following sums in the four times table. Numbers can go up, down, to the left, to the right and diagonally in all directions. Keep a tally chart. Two examples have been started for you.

x 4 sum	Tally
1 x 4	
2 x 4	
3 x 4	
4 x 4	
5 x 4	I
6 x 4	I
7 x 4	
8 x 4	
9 x 4	
10 x 4	

4	1	6	2	4	3	2	1	8	3
0	8	1	4	6	3	2	6	2	2
2	3	1	2	4	2	7	8	3	0
3	6	7	8	0	6	8	2	0	6
1	4	8	2	2	2	3	9	3	2
2	3	6	1	6	7	4	0	1	6
8	8	2	3	2	3	2	4	2	6
1	6	3	9	2	8	0	7	1	2
2	8	6	3	4	0	4	3	2	8
4	2	3	6	1	2	0	2	8	4

Race!

Name _____ **Date** _____

Use these two sets of sums to race with a friend. The same sums are in each set but in a different order. (Or you could time yourself with the first set and see if you can beat your own record with the second set.)

5 x 4 =	4 x 4 =
2 x 4 =	4 x 2 =
3 x 4 =	7 x 4 =
4 x 4 =	1 x 4 =
4 x 6 =	4 x 9 =
4 x 2 =	8 x 4 =
10 x 4 =	4 x 3 =
4 x 1 =	10 x 4 =
4 x 2 =	4 x 8 =
6 x 4 =	4 x 10 =
4 x 3 =	4 x 1 =
1 x 4 =	2 x 4 =
4 x 5 =	9 x 4 =
7 x 4 =	4 x 6 =
4 x 4 =	4 x 4 =
4 x 9 =	4 x 5 =
8 x 4 =	6 x 4 =
9 x 4 =	3 x 4 =
4 x 7 =	4 x 7 =
4 x 10 =	5 x 4 =

The 6 x Table

1 x 6 = 6

Left leg, left leg, left leg, right leg, right leg, right leg.

I have 18 fingers I am the best tickler I know!

2 x 6 = 12

| 2 | 4 | 6 | 8 | 10 | 12 |

3 x 6 = 18

I love 30!

4 x 6 = 24

5 x 6 = 30

30 30 30 30 30 30 30

6 x 6 = 36

I used to think 6 x 6 was 66.

Brilliant Ideas for Times Tables 7–9 © Molly Potter 2012

Join the sum to its answer

x6

Name _____ **Date** _____

Use colouring pencils to join each sum to its answer. One has been done for you.

1 x 6   **36**   **10 x 6**   **6 x 6**

7 x 6   **6**   **42**   **60**

4 x 6   **9 x 6**   **24**

12   **54**   **2 x 6**   **30**

5 x 6   **48**

8 x 6   **18**   **3 x 6**

Now answer these sums

5 x 6 =	3 x 6 =	6 x 3 =	6 x 9 =
2 x 6 =	10 x 6 =	1 x 6 =	8 x 6 =
6 x 2 =	6 x 1 =	6 x 5 =	9 x 6 =
4 x 6 =	6 x 2 =	7 x 6 =	6 x 7 =
6 x 6 =	6 x 6 =	6 x 4 =	6 x 10 =

Match it

Name _____ **Date** _____

Match each multiplication sum to its answer. Use colours and patterns to make each sum look the same as its answer. One has been done for you.

1 X 6
2 X 6
3 X 6
4 X 6
5 X 6
6 X 6
7 X 6
8 X 6
9 X 6
10 X 6
11 X 6
12 X 6

12   72   42

54   30

36   6   30

24   54

66

12   42

72

48

60   24   18

66

6   48

18   60   36

Brilliant Ideas for Times Tables 7–9 © Molly Potter 2012

Colour the arrows

Name _____ **Date** _____

Work out the sums and find the answer on the arrows. Colour all the arrows that have the wrong answers for the sum and the arrow that is left (the one with the correct answer), points to the next sum that you need to do. One has been done for you.

Which is the right answer?

Name _____ **Date** _____

Choose the correct answer for each sum and circle it.

1 x 6 = 4 6 18 20 12 10

6 x 4 = 12 20 48 24 56

2 x 6 = 6 12 24 8 18

6 x 9 = 49 63 54 42 56

3 x 6 = 24 12 18 60 15

6 x 7 = 56 63 54 42 44

4 x 6 = 16 24 30 18 28

6 x 2 = 2 6 12 4 10 14

5 x 6 = 12 11 36 10 30

6 x 5 = 25 60 20 30 40

6 x 6 = 30 36 46 12 40

6 x 10 = 35 50 40 40 60

7 x 6 = 38 36 45 30 42

6 x 1 = 4 5 12 10 6

8 x 6 = 38 48 36 41

6 x 6 = 16 36 12 24

9 x 6 = 90 32 63 54 66

6 x 8 = 54 36 56 42 48 12

10 x 6 = 40 60 10 70

6 x 3 = 6 12 24 10 18

Angry aliens

Name _____ **Date** _____

Work out why these aliens are angry by answering the multiplication sum to find the missing words.

6	12	18	24	30	36	42	48	54	60
golf	brain	eyeball	lollipop	human	hand	Saturn	bus	leg	nostril

He stole my
_____
(3 x 6)

He stole my fourth
_____ (9 x 6)
and played
_____ (1 x 6)
with it.

He broke
_____
(7 x 6)

He said I looked like a big
_____
(4 x 6)
and then licked me.
Yuk!

He really scared me by pretending to be a
_____
(5 x 6)

He bit off my
_____
(2 x 6)
probe.

He stuck a
_____
(8 x 6)
up my right
_____
(10 x 6)

He sat on my right
_____
(6 x 6)

Maze

Name _____ **Date** _____

Use different colours to join each multiplication sum to its answer in the maze.

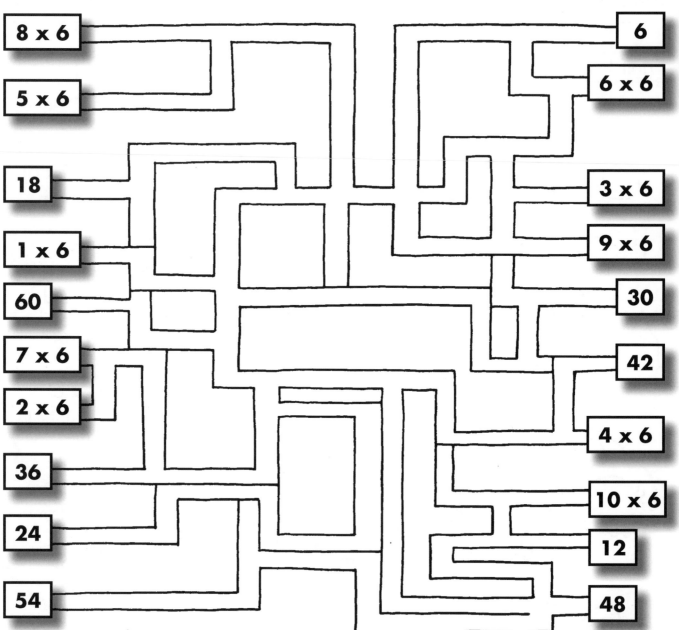

Now answer these:

1 x 6 = [   ]   8 x 6 = [   ]   6 x 2 = [   ]   6 x 3 = [   ]

7 x 6 = [   ]   6 x 6 = [   ]   6 x 4 = [   ]   6 x 5 = [   ]

5 x 6 = [   ]   10 x 6 = [   ]   6 x 9 = [   ]   6 x 6 = [   ]

4 x 6 = [   ]   9 x 6 = [   ]   6 x 1 = [   ]   6 x 10 = [   ]

3 x 6 = [   ]   2 x 6 = [   ]   6 x 7 = [   ]   6 x 8 = [   ]

Reveal the words

Name _____ **Date** _____

In each of the blobs below you will find one multiplication sum and four answers. You need to colour in all the **incorrect** answers in each blob and a word will form. One blob has been done for you.

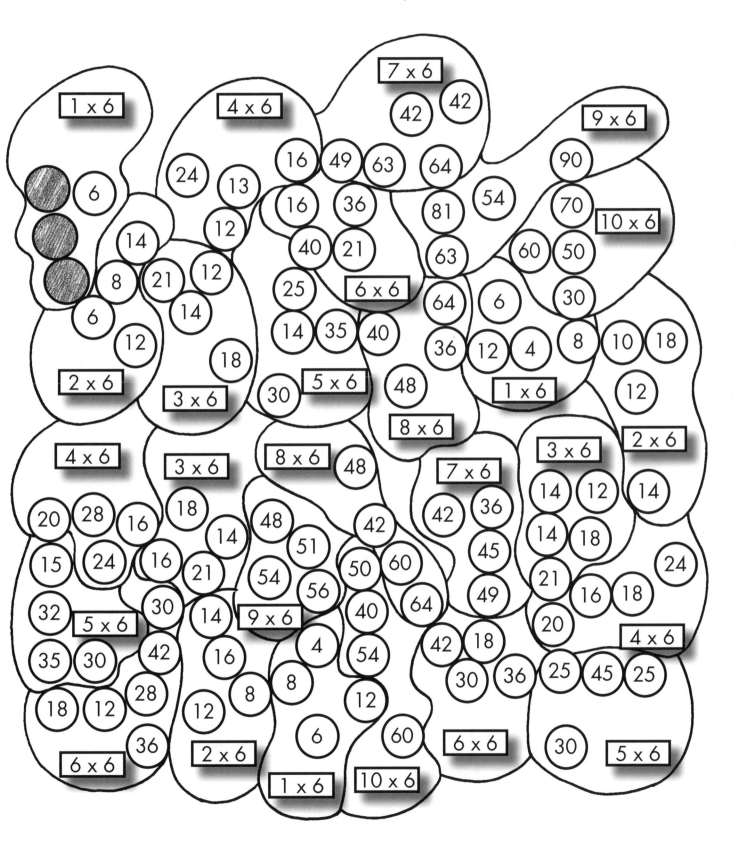

The most answers

x6 **Name** _____ **Date** _____

Tally the answers to the sums to find out which sum has its answer written the most times. Cross out the answers after you have counted them. One has been started for you.

x 6 multiplication sum	Number of answers
1 x 6	
2 x 6	I
3 x 6	
4 x 6	
5 x 6	
6 x 6	
7 x 6	
8 x 6	
9 x 6	
10 x 6	

30 42 42 24 18
48 60 54 42
60 36 18 54 24
42 24 42 36 18
18 48 12
54 30 48 6 36 18
42 42 54 42
42 24 54 6 48 24
36 48 30 36 18 48
24 54 24 42 60
54 12 18 42
24 54 48 24 54

x6   **Name** _____   **Date** _____

Use this de-coder to find out about Clive the Robot!

0	6	12	18	24	30	36	42	48	54	60
r	s	t	i	m	p	e	a	n	l	g

What does Clive eat?

_____ _____ _____ _____ _____ _____ _____
(6 x 4)   (7 x 6)   (10 x 6)   (6 x 8)   (6 x 6)   (2 x 6)   (1 x 6)

What is Clive's best friend?

_____   _____ _____ _____ _____ _____ _____ _____
(7 x 6)   (6 x 1)   (5 x 6)   (7 x 6)   (6 x 8)   (8 x 6)   (6 x 6)   (0 x 6)

What is the name of Clive's favourite restaurant?

_____ _____ _____ _____ _____
(6 x 4)   (6 x 6)   (6 x 2)   (6 x 7)   (9 x 6)

_____ _____ _____ _____ _____
(4 x 6)   (6 x 6)   (7 x 6)   (9 x 6)   (1 x 6)

What does Clive do when he sees a wheelbarrow?

_____ _____ _____ _____ _____ _____
(1 x 6)   (6 x 4)   (3 x 6)   (9 x 6)   (6 x 6)   (6 x 1)

_____ _____   _____ _____
(7 x 6)   (2 x 6)   (6 x 3)   (6 x 2)

Guess then check

Name _____   **Date** _____

Work out these sums then follow the line to check your answer.

3 x 9 = ☐   ☐ = 10 x 10

7 x 4 = ☐   ☐ = 4 x 5

5 x 7 = ☐   ☐ = 4 x 3

2 x 9 = ☐   ☐ = 3 x 6

3 x 7 = ☐   ☐ = 3 x 3

4 x 8 = ☐   ☐ = 4 x 6

6 x 5 = ☐   ☐ = 2 x 2

10 x 8 = ☐   ☐ = 3 x 5

2 x 7 = ☐   ☐ = 5 x 5

5 x 8 = ☐   ☐ = 6 x 7

4 x 9 = ☐   ☐ = 8 x 3

9 x 5 = ☐   ☐ = 2 x 6

6 x 6 = ☐   ☐ = 10 x 7

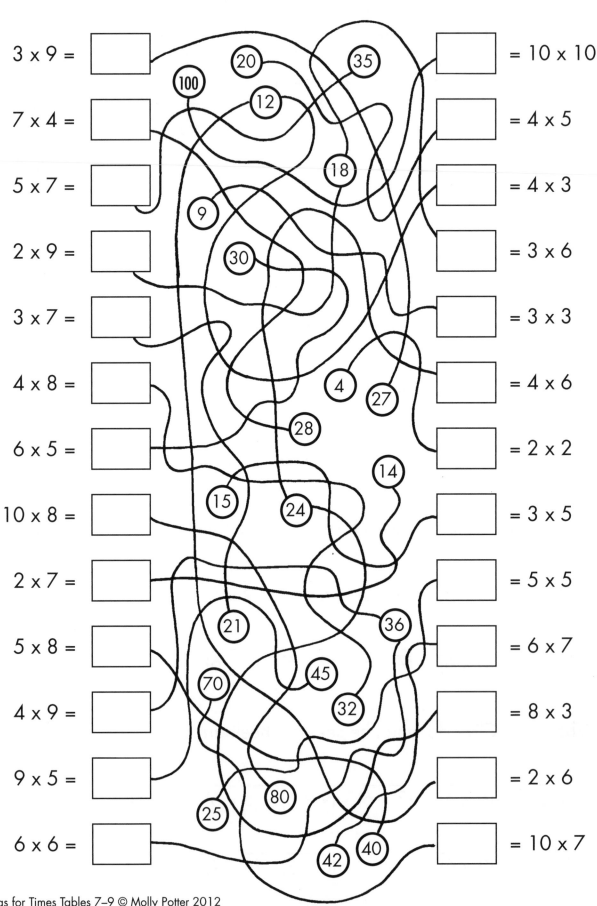

Multiply by?

Name _____ **Date** _____

Multiply the numbers in each table by the number above it and put the answer in the space below. One has been done for you.

x 2

7	2	5	4	1	9	8	3	6	10
14									

x 5

10	7	2	8	5	1	9	6	4	3

x 3

3	1	4	8	10	9	5	2	7	6

x 4

1	5	8	2	6	9	3	7	10	4

x 6

8	4	9	10	7	2	3	5	6	1

Snap or not snap

Name _____   **Date** _____

Look at each pair of sums. If the two sums come to the same amount –
tick the word snap ✓. If the numbers do not come to the same amount,
cross out the word ~~snap~~. The first one has been done for you.

1) 2 x 2 and 1 x 4

snap ✓

2) 2 x 3 and 1 x 6

snap

3) 5 x 3 and 2 x 8

snap

4) 3 x 4 and 2 x 6

snap

5) 5 x 5 and 3 x 8

snap

6) 4 x 4 and 3 x 5

snap

7) 2 x 8 and 4 x 4

snap

8) 2 x 10 and 4 x 5

snap

9) 3 x 6 and 4 x 4

snap

10) 2 x 4 and 1 x 8

snap

11) 2 x 5 and 1 x 10

snap

12) 8 x 5 and 4 x 10

snap

13) 3 x 3 or 1 x 10

snap

14) 6 x 4 and 3 x 8

snap

15) 3 x 7 and 7 x 3

snap

16) 3 x 8 and 5 x 5

snap

17) 3 x 6 and 2 x 9

snap

18) 4 x 9 and 6 x 6

snap

19) 5 x 6 and 4 x 7

snap

20) 3 x 10 and 5 x 6

snap

x2
x3
x4
x5
x6
x10

Dot to dot

Name _____ **Date** _____

First work out the answers to these sums. Next join the answers starting with the smallest number and finishing with the largest, then complete the shape by joining this back to your first dot. What shape do you make?

6 x 4 = ☐

2 x 2 = ☐

3 x 7 = ☐

3 x 3 = ☐

5 x 5 = ☐

5 x 4 = ☐   5 x 2 = ☐

3 x 5 = ☐

6 x 3 = ☐

3 x 4 = ☐

4 x 4 = ☐

2 x 7 = ☐

Which sum is wrong?

x2
x3
x4
x5
x6
x10

Name _____ **Date** _____

Look at these groups of sums. One of the sums does not come to the number in the box. Cross out the wrong sums. The first one has been done for you.

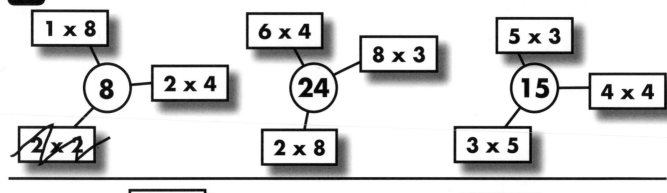

1 x 8
8
2 x 4
~~2 x 7~~ ✓

6 x 4
24
8 x 3
2 x 8

5 x 3
15
4 x 4
3 x 5

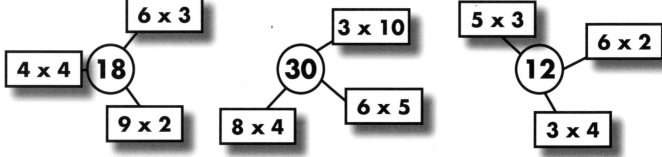

6 x 3
4 x 4   18
9 x 2

3 x 10
30
8 x 4   6 x 5

5 x 3
12   6 x 2
3 x 4

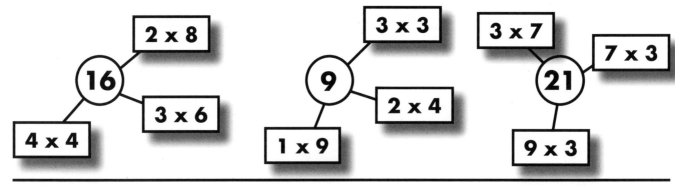

2 x 8
16
4 x 4   3 x 6

3 x 3
9
1 x 9   2 x 4

3 x 7
21   7 x 3
9 x 3

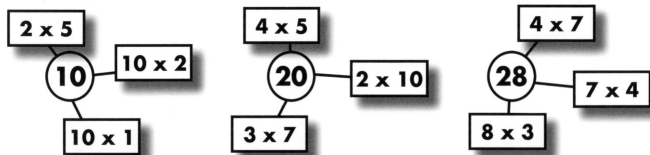

2 x 5
10   10 x 2
10 x 1

4 x 5
20   2 x 10
3 x 7

4 x 7
28   7 x 4
8 x 3

Now write two multiplication sums that come to the amounts in these boxes.

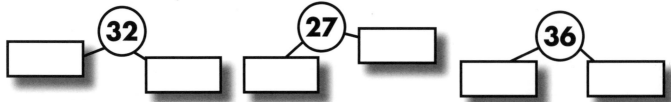

32   27   36

Top times

Name _____ **Date** _____

(1) Photocopy all 24 cards and laminate them if possible. Colour in the characters before you start! Share the cards equally between the players.

(2) Play one card at a time; always the top card.

(3) The youngest player starts by choosing a category (age, speed, height or magical power) to call. The winner is the person with the highest score (the answer to the multiplication sum) in that category. They win all the cards in that round and then they are the caller for the next round.

(4) If there is a tie, place the cards in play on the ground in a pile, and play the next card with the same caller. The winner of the next card takes the first pile too.

Doosy		**Gillom**		**Tecky**	
Age	(2 x 8)	Age	(6 x 5)	Age	(2 x 9)
Speed	(5 x 4)	Speed	(7 x 6)	Speed	(5 x 7)
Height	(6 x 6)	Height	(3 x 3)	Height	(9 x 5)
Magical Power	(9 x 4)	Magical Power	(5 x 10)	Magical Power	(6 x 4)
Groof		**Ooob**		**Jymi**	
Age	(3 x 3)	Age	(4 x 8)	Age	(3 x 2)
Speed	(3 x 7)	Speed	(3 x 8)	Speed	(4 x 4)
Height	(4 x 6)	Height	(3 x 2)	Height	.(4 x 6)
Magical Power	(4 x 4)	Magical Power	(6 x 3)	Magical Power	(10 x 10)

Top times

Harris

Age	(8 x 10)
Speed	(7 x 4)
Height	(9 x 5)
Magical Power	(3 x 2)

Mackil

Age	(6 x 2)
Speed	(6 x 4)
Height	(6 x 6)
Magical Power	(9 x 4)

Liffy

Age	(10 x 10)
Speed	(2 x 2)
Height	(7 x 6)
Magical Power	(3 x 8)

Doody

Age	(2 x 8)
Speed	(7 x 10)
Height	(6 x 6)
Magical Power	(5 x 2)

Canila

Age	(10 x 4)
Speed	(5 x 4)
Height	6 x 6)
Magical Power	(9 x 4)

Hoff

Age	(8 x 4)
Speed	(6 x 10)
Height	(3 x 3)
Magical Power	(9 x 4)

Imillie

Age	(4 x 2)
Speed	(7 x 3)
Height	(6 x 6)
Magical Power	(5 x 5)

XR7

Age	(10 x 5)
Speed	(8 x 3)
Height	(2 x 6)
Magical Power	(5 x 6)

Drawk

Age	(9 x 5)
Speed	(8 x 6)
Height	(3 x 5)
Magical Power	(9 x 2)

Top times

Hoddy

Age	(5 x 3)
Speed	(1 x 3)
Height	(2 x 4)
Magical Power	(9 x 3)

Clive

Age	(10 x 3)
Speed	(4 x 4)
Height	(2 x 10)
Magical Power	(2 x 5)

Grolp

Age	(7 x 4)
Speed	(8 x 5)
Height	(10 x 6)
Magical Power	(2 x 3)

Wooth

Age	(7 x 5)
Speed	(1 x 4)
Height	(4 x 3)
Magical Power	(7 x 2)

Krik

Age	(4 x 5)
Speed	(6 x 4)
Height	(9 x 6)
Magical Power	(3 x 6)

Juliap

Age	(4 x 10)
Speed	(6 x 3)
Height	(3 x 4)
Magical Power	(4 x 6)

Mr Tog

Age	(6 x 4)
Speed	(7 x 10)
Height	(3 x 4)
Magical Power	(6 x 6)

Shantilly

Age	(4 x 4)
Speed	(8 x 4)
Height	(2 x 9)
Magical Power	(10 x 3)

Dave

Age	(3 x 5)
Speed	(3 x 3)
Height	(9 x 4)
Magical Power	(6 x 3)

Card collector

x2
x3
x4
x5
x6
x10

This simple game revises the 2, 3, 4, 5, 6 and 10 times tables. You need at least two players, a dice, a counter for each player, the game board on page 60 enlarged to A3 size and the cards (pages 58 and 59) cut up and shuffled.

To play:

① Place your counters on the start, throw the dice and move that number of squares.

② If you land on a TT space, somebody else picks up a Times Table question and asks you it. If you get the answer correct (the answer is in the corner of the card) you keep the card. If you get it wrong, you do not keep the card and it is the next person's turn. If you land on an arrow, move in the direction of the arrow on your next go. If you land on a square with just a picture, that is the end of your go.

③ Take it in turns to have a go. The game stops when the first player crosses the finishing line but the winner is the person who has won the most cards at that point.

④ In the unlikely event of all the cards being used up, count how many cards each player has, keep their score and then return the cards to a pile, turn the cards over and reuse them.

1 x 2 = 2	1 x 3 = 3	1 x 4 = 4
2 x 2 = 4	2 x 3 = 6	2 x 4 = 8
3 x 2 = 6	3 x 3 = 9	4 x 3 = 12
2 x 4 = 8	3 x 4 = 12	4 x 4 = 16
2 x 5 = 10	3 x 5 = 15	4 x 5 = 20
2 x 6 = 12	3 x 6 = 18	4 x 6 = 24

7 x 2 = ☐ 14	7 x 3 = ☐ 21	7 x 4 = ☐ 28
8 x 2 = ☐ 16	8 x 3 = ☐ 24	8 x 4 = ☐ 32
2 x 9 = ☐ 18	3 x 9 = ☐ 27	4 x 9 = ☐ 36
10 x 2 = ☐ 20	10 x 3 = ☐ 30	10 x 4 = ☐ 40
1 x 5 = ☐ 5	1 x 6 = ☐ 6	1 x 10 = ☐ 10
2 x 5 = ☐ 10	2 x 6 = ☐ 12	2 x 10 = ☐ 20
3 x 5 = ☐ 15	3 x 6 = ☐ 18	3 x 10 = ☐ 30
5 x 4 = ☐ 20	6 x 4 = ☐ 24	4 x 10 = ☐ 40
5 x 5 = ☐ 25	6 x 5 = ☐ 30	10 x 5 = ☐ 50
5 x 6 = ☐ 30	6 x 6 = ☐ 36	10 x 6 = ☐ 60
7 x 5 = ☐ 35	7 x 6 = ☐ 42	7 x 10 = ☐ 70
8 x 5 = ☐ 40	8 x 6 = ☐ 48	8 x 10 = ☐ 80
5 x 9 = ☐ 45	6 x 9 = ☐ 54	10 x 9 = ☐ 90
10 x 5 = ☐ 50	10 x 6 = ☐ 60	10 x 10 = ☐ 100

FINISH

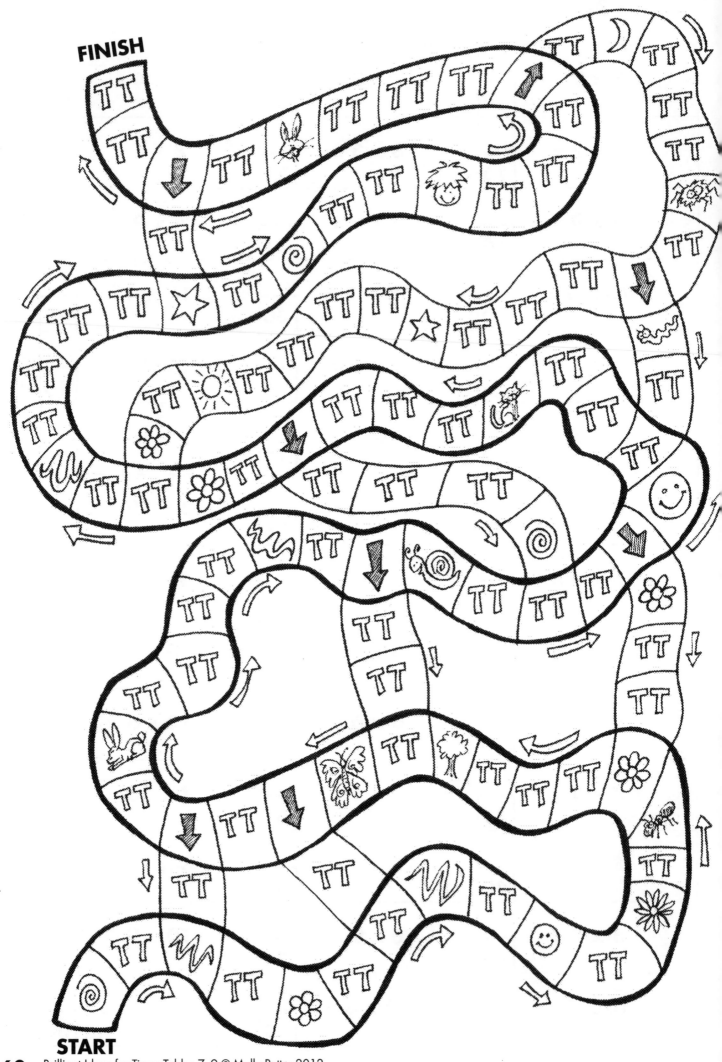

START

Factors, products & multiples

Name _____   **Date** _____

FACTORS

Consider the number 24.

All of these multiplication sums come to 24:

$6 \times 4 = 24$

$2 \times 12 = 24$

$1 \times 24 = 24$

$8 \times 3 = 24$

So all these number, 1, 2, 3, 4, 6, 8, 12 are said to be FACTORS of 24.

PRODUCTS

If we multiply 3 and 8 together. The answer is called the PRODUCT of 8 and 3.

The PRODUCT of 6 and 4 is also 24.

MULTIPLES

The MULTIPLES of 3 are all the numbers in the three times table and beyond.

So MULTIPLES of 3 include: 3, 6, 9, 12, 15, 18, 21, 24, 27, 30, 33, 36, 39.......

So 24 is a MULTIPLE of 3.

Now answer these questions.

1. Is 3 a factor of 12?   Yes/No

2. Is 21 the product of 7 and 3?
                                    Yes/No

3. Is 40 a multiple of 10?   Yes/No

4. Write two factors of 18.
   _____ and _____

5. What is the product of 5 and 3?

   _____

6. Write two multiples of 5
   _____ and _____

7. Is 5 a factor of 12?   Yes/No

8. What is the product of 4 and 7?

   _____

9. Is 20 a multiple of 4?   Yes/No

10. Is 4 a factor of 16?   Yes/No

11. Is 32 the factor of 4 and 8?
                                    Yes/No

12. Is 48 a multiple of 6?   Yes/No

13. What is the product of 7 and 6?

    _____

14. Is 36 the product of 6 and 6?
                                    Yes/No

Answers

Answers are not given for activities where these are simply
the result of a multiplication sum.

Activity sheet 2:
Colour by numbers

(2, 5 and 10 times tables)
It makes a seahorse.

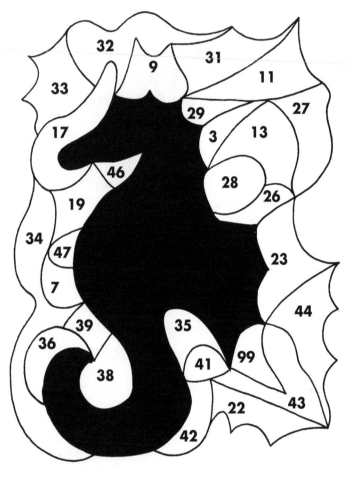

Activity sheet 3:
Which is closer?

(2, 5 and 10 times tables)

- 2 x 8
- 9 x 5
- 7 x 5
- 9 x 2
- 10 x 10
- 8 x 5
- 5 x 10
- 2 x 7
- 10 x 7
- 2 x 3
- 3 x 5
- 9 x 10

Activity sheet 4:
Which sum has the most answers?

(2, 5 and 10 times tables)

sum (2 x)	No.	sum (5 x)	No.	sum (10 x)	No
2 x 3	1	5 x 3	4	10 x 4	3
2 x 4	2	5 x 4	2	10 x 7	2
2 x 5	2	5 x 5	2	10 x 9	2
2 x 6	3	5 x 6	3	10 x 10	2
2 x 7	5	5 x 7	5		
2 x 8	3	5 x 9	4		
2 x 9	6				

2 x 9 has the most answers.

Activity sheet 5:
Find out about Snook

(2, 5 and 10 times tables)

- brick stew
- dustbin juice
- the moon
- in a sock
- traffic lights

Activity sheet 7:
Dot to dot

(2, 5 and 10 times tables)
It says 'YOU STAR'.

Activity sheet 9:
How many?

(2, 5 and 10 times tables)

1) 2 x 8 = 16
2) 4 x 2 = 8
3) 6 x 10 = 60
4) 5 x 5 = 25
5) 2 x 7 = 14
6) 9 x 5 = 45
7) 5 x 3 = 15
8) 2 x 5 = 10
9) 8 x 10 = 80
10) 3 x 10 = 30
11) 6 x 10 = 60
12) 8 x 5 = 40
13) 5 x 7 = 35
14) 5 x 4 = 20

Activity sheet 11:
Get rid of the wrong answers
(3 times table)

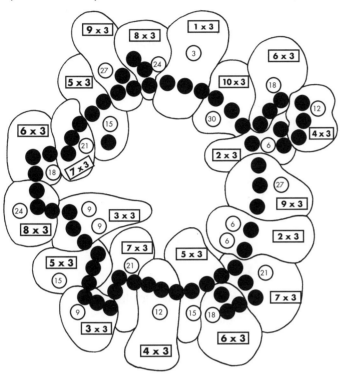

Activity sheet 17:
Which is closer?
(3 times table)

- 1 x 3
- 2 x 3
- 8 x 3
- 5 x 3

- 7 x 3
- 9 x 3
- 4 x 3
- 1 x 3

- 4 x 3
- 7 x 3
- 6 x 3
- 3 x 3

Activity sheet 18:
Find out about Jibby
(3 times table)

- hats
- a sheep
- boys
- by space hopper
- Pattobot

Activity sheet 19:
Closest to the target
(3 times table)

3 x 6 gets closest to the target.

Activity sheet 24:
Hidden picture
(4 times table)

The picture is a dragon.

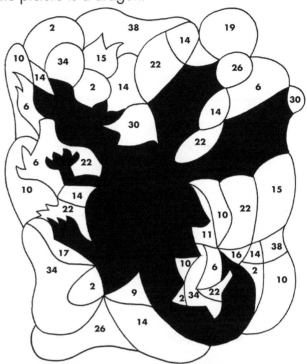

Activity sheet 27:
Which is closer?
(4 times table)

- 5 x 4
- 7 x 4
- 3 x 4
- 7 x 4

- 2 x 4
- 1 x 4
- 4 x 4
- 2 x 4

- 7 x 4
- 8 x 4
- 7 x 4
- 10 x 4

Activity sheet 36:
Reveal the words
(6 times table)

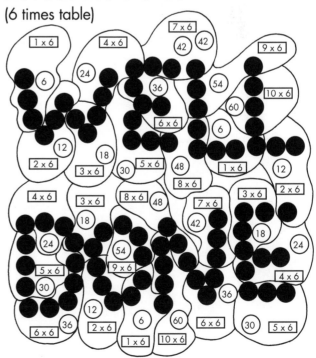

Activity sheet 37:
The most answers
(6 times table)

x 6	Number of answers
1 x 6	2
2 x 6	2
3 x 6	7
4 x 6	9
5 x 6	3
6 x 6	5
7 x 6	12
8 x 6	7
9 x 6	9
10 x 6	3

7 x 6 has the most answers.

Activity sheet 38:
Find out about Clive the Robot
(6 times table)

- magnets
- a spanner
- Metal Meals
- smiles at it

Activity sheet 41:
Snap or not snap
(2, 3, 4, 5, 6 and 10 times tables)

1) snap
2) snap
3) not snap
4) snap
5) not snap
6) not snap
7) snap
8) snap
9) not snap
10) snap

11) snap
12) snap
13) not snap
14) snap
15) snap
16) not snap
17) snap
18) snap
19) not snap
20) snap

Activity sheet 42:
Dot to dot
(2, 3, 4, 5, 6 and 10 times tables)
The shape is a cross

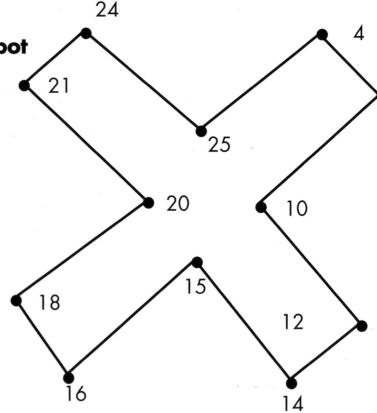